Dan's Practical Guide to

Least Toxic

Home Pest Control

DAN'S PRACTICAL GUIDE TO

LEAST TOXIC
HOME PEST CONTROL

Carpenter Ants ♦ Carpet Beetles
Clothes Moths ♦ Cockroaches
Fleas ♦ Fruit Flies
Head Lice ♦ Houseflies
Indian Meal Moths
Mice & Rats
Silverfish ♦ Spiders
Sugar Ants ♦ Termites
Wasps & Yellow Jackets

by Dan Stein, M.S.

HULOGOSI

Illustrated by JOHN F. GRAHAM

Book design by Alan Trist
Published by Hulogosi Communications, Inc.
P.O. Box 1188, Eugene, Oregon 97440
Printed in the U.S.A.

FIRST EDITION

Library of Congress Cataloging-in-Publication Data

Stein, Dan, 1954 –
Least toxic home pest control / by Dan Stein;
[illustrated by John F. Graham].
p. cm.
ISBN 0-938493-15-9
1. Household pests — Control. I. Title.
TX325.S74 1991
648'.7–dc20   91–12305
                                                    CIP

DISCLAIMER

While every effort has been made to select safe, least toxic methods for you to use, the author cannot accept any responsibility for accidents or problems that may occur while using any of the methods recommended in this book. If you have any reservations about doing your own pest control, hire a professional to do it for you.

Reference to commercial products and their trade names (in bold type) is made for the purposes of clarity and no discrimination of other companies with similar products is intended.

CONTENTS

Introduction   vii

BASICS OF LEAST TOXIC PEST CONTROL

What is Least Toxic Pest Control?   1
Straight Talk About Chemicals   2
Is Non-Chemical Pest Control Really Possible?   3
Non-toxic Pest Control Methods and Materials   5
Pesticide Use and Safety   7
Understanding Insect Life Cycles   10

15 MOST COMMON HOUSEHOLD PESTS

Carpenter Ants   11
Carpet Beetles   17
Clothes Moths   21
Cockroaches   23
Fleas   27
Fruit Flies   33
Head Lice   37
Houseflies   41
Indian Meal Moths   45
Mice and Rats   49
Silverfish   53
Spiders   57
Sugar Ants   61
Termites   67
Wasps and Yellow Jackets   73

APPENDICES

Other pests   79
Close Encounters of the Fourth Kind – Exterminators   81
Least Toxic Lawn Care   83
Sources of Least Toxic Products   84
For More Information...   86

This book is for those folks who want to learn how to do the most with the least. It is also for my friends in the insect world in the hope that the people who read this book will learn to accept more of your kind among their kind.

INTRODUCTION

This book will help you solve most of the common household pest problems with an absolute minimum amount of insecticides. The secret to controlling pests without insecticides is *out-smarting them*. This is not always as easy as it may sound. While humans have an undeniable edge in brain size, insects have honed excellent survival skills through millions of years of evolution that tend to even out the competition. The challenge is, can humans come out on top, without polluting our homes. We can do it!

Many of the pest control methods in this book rely heavily on common sense. Try not to let this make you feel that they must be too simple to work. Give these ideas a fair shake and you will be pleasantly surprised how effective they are.

But common sense is only half the story. In recent years new, less toxic insecticides, growth regulators, pheromones, and trapping methods have become available that are changing the pest control industry. Some people are predicting that in a few years we may be using no toxicants (poisons) at all but only biological tools and methods to control pests. This direction away from toxicants is pretty exciting!

A few words about the scope of this book. As I, the author, have worked only with the pests found in the Pacific Northwest, this book is inevitably oriented to this region. The species mentioned are those found in the Northwest. Other regions will have different species but they will still be the same types of pests. The only serious limitation to this approach is that there may be pests from other regions not covered at all here. I don't expect there will be many though. Happy Hunting!

BASICS OF
LEAST TOXIC PEST CONTROL

WHAT IS LEAST TOXIC PEST CONTROL?

The concept of 'least toxic pest' control is one I invented* to help explain IPM, Integrated Pest Management, to my pest control customers. (I run a pest control and consulting company that uses IPM methods.) Just so you know, IPM is a decision-making process that looks at pest control within the context of the shared pest and human ecosystem. The formal IPM process has five parts:

1. Identify the pest and learn about its biology.

2. Monitor the pest to determine how serious a problem it is, and determine at what point action needs to be taken (Action Threshold).

3. Explore all the control options, including non-chemical methods; list each method's pros and cons.

4. Select a control strategy and make your least toxic treatments.

5. Evaluate how your method worked and modify it if necessary.

While I use this process to develop new methods as part of my work, I realized that for many people it was too abstract. Over time I realized that the essence of IPM was simply using the best combination of 'least toxic' methods and materials available. Most

---

* OK, the truth is, like many great ideas whose time has come, different people started talking about 'least toxic' pest control at about the same time. I just like to think I came up with it first.

THE BUG FILE ♦ The Japanese character for ant is a combination of the

1

of the time now I skip the lectures on IPM and talk instead about 'least toxic' pest control.

Please note that an IPM program (or a '*least* toxic' pest control program) is committed to 'least toxic' methods, but not necessarily '*non*-toxic' methods. Pesticides *can* be used as part of IPM programs unless you decide not to include any in your treatment options. In any case, IPM uses alternatives to pesticides whenever possible. If pesticides are determined to be necessary, IPM always uses the least toxic pesticide in the smallest effective amount.

Many people are not aware that organic food can legally be grown with the use of certain specific 'safe' pesticides. Thus even 'organically grown' food is really grown under IPM methods.

For the reader of this book, the IPM (or *least toxic*) process can be summarized in three words:

STOP AND THINK!

Take the time to figure out what is going on in your home before taking any action. Then, decide on your best pest control strategy, with the help of this book of course!

STRAIGHT TALK ABOUT CHEMICALS

I have to admit that I am a little defensive about recommending the use of chemicals and pesticides of any sort and want to clear the air about this. Some people think I should be using and recommending *no* chemicals. Others feel my approach is not chemically intensive enough to possibly work. My response:

First off, everything is made of chemicals. Chemicals are simply molecules attached together in different ways and are inherently neither good nor bad. Many things determine whether

---

characters for "insect", "unselfishness", "justice" and "courtesy". ♦ Dragonflies

a chemical is 'good' or 'bad' including its short- and long-term toxicity, its persistence, its breakdown products, and effects on non-target organisms. Without question, certain chemicals do put our own health and the planet's health at risk. Unfortunately, often we do not know for sure (or at all!) which are the good ones and which are not.

My solution to this uncertainty is to cut down as much as possible on *all* unnecessary chemical use, *and* at the same time, to decide what chemicals I am willing to continue using despite possible risks. All chemicals have their risks, including not just pesticides, but tobacco, caffeine, alcohol, gasoline, household cleansers, smoke from wood stoves, fast foods, etc... All need to be looked at.

The chemical methods suggested in this book are what I believe to be the safest ones that will get the job done. At the risk of alienating all my readers, I offer both chemical and non-chemical methods so people can choose a method that won't get them into trouble with their peers.

IS NON-CHEMICAL PEST CONTROL REALLY POSSIBLE?

Of course it is, *but* you may have to get used to higher levels of pests in your house. This is not a bad thing. The truth is we have grown accustomed to a level of pest control that is unrealistic and get unreasonably upset by a very low level of pests.

We must learn to live with a few pests in our homes. The term 'pest', like the term 'weed', refers to an organism 'out of place'. We must learn to have more tolerance for non-humans of all kinds, even in our homes. I try to think of a few pests in my house as just a few more poorly trained pets. (That's a joke.)

have existed for at least 50 million years. Early dragonflies had wing spans over

Theoretically, here's how pest control works:

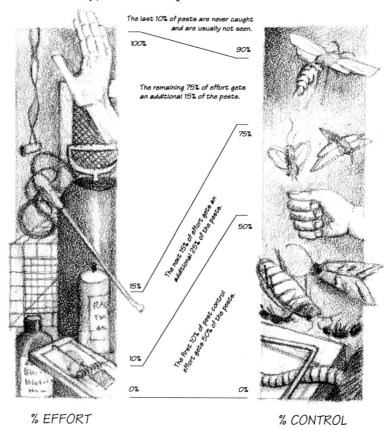

The last 10% of pests are never caught and are usually not seen.

The remaining 75% of effort gets an additional 15% of the pests.

The next 15% of effort gets an additional 25% of the pests.

The first 10% of pest control effort gets 50% of the pests.

100%   90%

75%

50%

15%

10%

0%   0%

% EFFORT   % CONTROL

Now here is the key...

The first 25 percent of pest control effort can and SHOULD be primarily NON-CHEMICAL which, in theory, will rid you of 3/4 of your pests.

If you can live with a few pests (now re-named pets) you can make a significant impact with non-chemical methods alone.

two feet long. ♦ A honeybee queen can lay more than a million eggs during her

NON-TOXIC PEST CONTROL
METHODS AND MATERIALS

The following list contains some non-toxic (or almost non-toxic) methods and materials for use in our pest control efforts. Note this list includes both concepts and products. While corporations not surprisingly prefer to promote products over concepts (easier to sell), these concepts are essential to the process of 'least toxic' pest control and need to be recognized and promoted.

■ **Sanitation.** Nothing is more important in 'least toxic' pest control than keeping your home clean. A clean house means no free lunch for the bugs. Train yourself not to leave food or dirty dishes out overnight. Try to clean the areas around and under stoves and refrigerators at least once every few years.

■ **Moisture Control.** Many insects are as attracted to moisture as they are food. Fix leaks and drips promptly. Try to keep the area around sinks wiped dry.

■ **Caulking.** Insects need homes too, and because they are small, they prefer small homes. Caulking cracks eliminates insect hiding places and is always worth the effort. Concentrate caulking efforts in the kitchen, bathroom, and around doors and windows.

■ **Screening.** Screening is a very important component of 'least toxic' pest control. It is especially useful in keeping out houseflies, mosquitoes, and night flying moths. Also make sure your crawl space vents are tightly screened to keep out rodents and possums.

■ **Traps.** Many kinds of traps are used in pest control including the following:

- Cockroach sticky traps
- Glueboards for catching mice (Yuck!)
- Pheromone traps

---

three year life span. ♦ Most insects have multi-faceted compound eyes: Housflies

5

- Flypapers of different kinds
- Mouse, rat, gopher, and mole traps
- Light Traps ('bug zappers') for flying insects (recommended for indoor use only)

■ **Insecticidal Soap.** Insecticidal soap is a specially formulated soap product used mostly against soft-bodied insects. Indoors, it can be used for flea control. Outdoors, it is useful against many ornamental and garden pests.

■ **IGRs.** Insect Growth Regulators, are man-made versions of the natural growth regulators found within insects. Not non-chemical, they are chemicals of low toxicity to humans and pets that prevent insects from reaching maturity so they can't reproduce.

■ **Pheromones.** Pheromones are insect scent hormones that are often used as bait in traps to trick unsuspecting males into thinking they have found a female. Each insect has a specific pheromone that attracts only it's own kind. Many pests can be monitored or controlled with pheromone traps.

■ **Repellents.** Repellents, not too surprisingly, repel pests. There are not many commercially available besides mosquito repellents, but some herbs have short term repellent properties (Citronella, Eucalyptus, Wormwood...). Also, deer repellents.

■ **Biological Control.** More commonly used in agriculture, biological control is the intentional use of 'good' organisms against 'bad' ones. It is not all that practical in the home at this point, but Gecko lizards are used in the South to control roaches (Their use has a couple drawbacks - they bark and bite!). Minute harmless Trichogramma wasps can be used to control meal moths. One common example of biological control is the use of cats to catch mice.

have about 4,000 facets, honeybee drones, 13,000 facets. ♦ Dung beetles were

■ **Home Remedies.** All sorts of home remedies exist. Some have been passed down for generations, and still don't work (i.e., cucumber peels to discourage roaches). Others work but have not been scientifically documented. As long as your home remedy does not involve the use of something poisonous, by all means, give it a try. You may stumble onto something great.

[Here is an offer for you frustrated home scientists. Send me your home remedies and your 'data' supporting your claims, and I will acknowledge any great ideas in the next edition of this book.]

PESTICIDE USE AND SAFETY

Not many people know that most insecticide poisoning takes place not on the farm, not at work, but at home. This is due primarily to kids getting into household chemicals and gardeners taking few safety precautions while applying insecticides. Even when using many of the 'organic' insecticides, there is a very real danger of poisoning yourself.

The key to not being a poisoning victim is to use common sense. If you stop and think – *safety* – you will avoid that unpleasant and expensive visit to your local emergency room.

Always read the pesticide label! The label on a pesticide container is a legal document that must be read before the product is used under penalty of law. Though few people go to jail for not reading labels, the intent of the law is good. A recent survey found that less than half of all people read the label before applying pesticides. Labels should be read *every time* you use a pesticide product.

Minimum safety equipment should include the following:

● Plastic or rubber gloves.
● Dust mask (when using dust materials).

worshipped by ancient Egyptians because of their habit of making round balls of

7

- Pesticide rated respirator (when using sprays). Good quality disposable respirators can be purchased for about $15. Despite being disposable, if kept dry they will last a long time.
- Long sleeved shirts.
- Long pants.
- Enclosed shoes.

After you are finished spraying, take a shower and put on fresh clothes.

Remember to store pesticides out of reach of children. Also, write the date of purchase on the product. Old pesticides as a rule are less effective. Rule of thumb – keep insecticides no more than one year.

THE THREE MAIN INSECT BODY PARTS

Head   Thorax   Abdomen   (technically a
                                     gaster when found
                                     on an ant)

dung that were rolled around, symbolizing the sun being pushed across the sky.

Non-target exposure to pesticides can be a serious problem
whenever pesticides are used.

Outside, unwanted spray may drift onto vegetable gardens
or neighbors' yards, as well as onto yourself.

Inside, people spend more time in close proximity to sprayed
areas and thus have more direct exposure to pesticides.
The process of "airing out" takes much longer.

♦ How fast can an insect fly? Houseflies fly at about 4 mph; Mosquitoes fly up

UNDERSTANDING INSECT LIFE CYCLES

Knowing a little bit about the biology of an insect is very useful in figuring out ways to control it. Armed with a little knowledge, you can determine at what stages the insect is most vulnerable. Most insects pass through several stages of metamorphosis where the insect takes on different body types.

Everyone is familiar with the butterfly that starts off as an egg that hatches into a caterpillar that wraps itself into a cocoon that emerges as a butterfly. All insects go through some variation of this process.

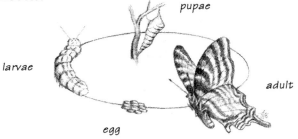

Insects with complete metamophosis pass through these four stages

When an insect passes through all four stages (technically called egg, larvae, pupae, and adult) this process is called *complete metamorphosis*. Ants, fleas, flies, yellow jackets and meal moths all pass through these four stages.

When an insect passes through three stages only (egg, nymph, adult), this is called *incomplete metamorphosis*. Roaches and silverfish both pass through these three stages.

Be aware that the adult stage is not always the 'pest' stage. Most of you know that kids can be pests too. Carpet beetles, clothes moths and Indian meal moths are all pests during immature stages.

---

to 5 mph; Honeybees can fly up to 13 mph; Dragonflies have been recorded at 40

CARPENTER ANTS
(Camponotus spp.)

Personally, I like carpenter ants. I am always impressed with how well organized their nests are, how hard they work, how they care for their young, and make good use of the resources around them, even as I close in to destroy them. I was pleasantly surprised that I was not alone in my affection for ants when I learned that the Japanese character for 'ant' is a combination of the symbols for 'insect', 'unselfishness', 'justice' and 'courtesy'.

DESCRIPTION

Carpenter ants are large, wood-infesting ants that are very common, some would say too common, throughout the Northwest and other parts of the country. Carpenter ants when observed in profile with a magnifying glass have two identifying characteristics - a smooth rounded 'back' with no humps, and a single tiny 'pyramid' (pedicel) at the 'waist'.

mph. ♦ Orthodox Jews when they start a boy's education have the boy kiss a drop

(You may be wondering how to capture an insect to identify it without squashing it? An easy way is to gently flick it into a small container that you can place in the freezer for a few minutes.)

Every region has its own species of carpenter ants. Two species are most common in western Oregon and Washington, the region where I live. The Modoc carpenter ant, *(Camponotus modoc)* is all black and usually between 1/4 and 1/2 inches long. The Bi-colored carpenter ant *(Camponotus vicinus)* is about the same size with a reddish colored mid-section (thorax). This species will not infrequently nest in the ground. Most other species nest in or next to wood (a house will do...).

The presence of carpenter ants within a home can be determined in several ways:

- Seeing the ants themselves inside the home over an extended period of time.

- Finding piles of coarse sawdust (called frass) that the workers push out of their nests, usually under the house.

- Hearing them rustle within a wall, especially at night.

LIFE CYCLE

Carpenter ants are social insects that generally nest in or near wood. Unlike termites that can digest the cellulose in wood (with the help of protozoa in their guts), carpenter ants cannot eat wood. Instead they eat insects and sometimes nectar. The damage they do to homes occurs when they make nests. Fortunately, they are fairly lazy creatures and often prefer wall voids and moisture-damaged wood to dry, sound wood. Nevertheless they will infest sound wood if they have a mind to.

A colony is started in two different ways. Either a winged

of honey placed on a book to teach him the sweetness of learning. ♦ In some areas

queen selects a new site on her own, or an existing colony splits into two or more nests creating 'satellite' nests.

The queen's only job is to lay eggs. The workers do everything else. Under normal circumstances the period from egg to adult takes about two months. In spring or summer, winged reproductives, both males and females, fly off to mate and establish new colonies.

RECOMMENDED CONTROL MEASURES

There are three steps in the 'least toxic' control of carpenter ants:

1. Locating the nests.
2. Treating the nests.
3. Making sure they don't come back.

1. LOCATING NESTS

The key to 'least toxic' carpenter ant control is locating the nest or nests. Once a nest is located, a 'spot' treatment can be made using a small amount of a low toxicity insecticide. There may be one or several nests in a home, and other nests outside. Usually an outside nest, in for example an old stump, will be the source of the nests within the building. Finding the nest is not always easy but makes for an interesting adventure. The most productive method for locating nests is to follow trails of foraging ants. These trails will include ants going in both directions.

One direction will be heading away from their nest to their 'feeding grounds'. You will know you are headed in this direction if the trail gradually fans out and disappears.

A trail headed towards a nest will remain well defined until it disappears into a crack in a building or a stump, or some other type of wood in which the ants are nesting. Many times they will be

cockroaches were thought to attack bed bugs and so were tolerated as the

carrying bug parts as they head towards their nest.

Often, though not always, the nest in a building will be near the crack through which they enter the building. Occasionally the nest may be quite some distance from their point of entry.

As carpenter ants are most active at night, you will have best results following trails if you look after dark, on evenings when the temperature is 60°or above. These trails follow the same routes throughout the summer so that if you get discouraged one night, you can pick up their trail where you left off on another evening.

2. TREATING NESTS

Locating the nests can take time; treating them is relatively easy. An outside nest can be destroyed with ten or more seconds of a pyrethrin aerosol (such as **Xclude***) sprayed directly into the nest site. To get to ants nesting within a wall void, an 1/8 inch hole can be drilled into the wall and an aerosol with an injection straw can be used to squirt a few seconds of chemical into and directly around the nest site. A few products (such as **Tri-Die** and **Revenge**) come with these straws.

In two weeks, check to see if the trails leading to the nest are still active. If they are, treat again.

Accessible nests of ants can be removed with a vacuum cleaner. Nests in wood can also be physically removed by a carpenter and the infested wood replaced. If the entire nest is removed, no spraying is necessary.

3. KEEPING THEM FROM COMING BACK

To some extent, carpenter ants can be discouraged from returning (or taking up residence in your home in the first place) by taking

---

* A list of all brand names mentioned in the text is found in the Appendix.

---

lesser of two evils. ♦ The cheerful "cricket on the hearth" was attracted to both

some preventative measures:

- Cracks that may allow ants easy access into your home should be caulked or otherwise sealed shut.

- Moisture-damaged wood, which attracts ants, should be removed from your home if possible, and the source of moisture should be eliminated.

- Stumps and other sources of wood debris can be removed from your yard. If this is not practical, stumps should be inspected on a routine basis for nests.

- Bark mulch should not be used adjacent to homes.

- Firewood should be stored away from the house.

- Tree branches should never be allowed to touch the house itself as this gives ants another access route.

- A clearance of at least six inches should be maintained between the bottom of wood siding and ground level to protect the siding from water damage and to make it harder for ants to 'find' the building. Don't allow the soil level to creep up around the house.

[A kit is available from the author for people interested in treating carpenter ants themselves but who want some professional assistance. The kit contains all the necessary 'least toxic' chemicals, safety equipment, a 16-page instruction booklet and two phone consultations. See Inside Back Cover of book for details.]

the warmth of the fireplace and the cracks in the masonry where it made its

CARPET BEETLES
(Dermestidae Family)

C arpet beetles lead dual lives, and even have two names. Many people are familiar with them indoors as a pest of woolens. But few know that the very same beetles when seen outdoors are called flower beetles, do no damage at all, and are admired for the beautiful patterns on their backs when looked at up close.

DESCRIPTION

Carpet beetles, while not all that familar to many homeowners, cause more damage as a group than the more notorious clothes (wool) moths. The adults are slightly smaller than lady beetles (lady bugs) and depending on the species are either, black, or mottled with white, yellow and brown patterns. The larvae are slightly smaller, brown, hairy and carrot shaped.

There are several species of carpet beetles found in homes including the black carpet beetle, the varied carpet beetle and the furniture carpet beetle. Their habits are all pretty similar so it really doesn't matter which one you have.

LIFE CYCLE

Most species have similar life cycles. The adults feed for about a month primarily on flower pollen. They enter homes through open windows. The larvae feed on a wide variey of foods including carpets, clothes, hair, leather, wasp and bird nests, and people food. Larvae feed and molt for up to several months, then pupate for up to two weeks. Adults are sometimes found on windows trying to get outside to continue their life cycle.

RECOMMENDED CONTROL MEASURES

Carpet Beetles (and clothes moths – see next chapter), can be particularly frustrating pests. Even when their damage is not widespread, a small hole or two is all it takes to ruin a favorite sweater or suit. They are also difficult to get rid of once and for all. While the control process does not take a lot (or any) chemicals, it does take a fair amount of effort.

■ **Don't let them get started.** This is one pest that is often easier to prevent than to control. Try to avoid bringing infested items into the house. Also, regular thorough cleaning helps prevent the buildup of beetle food and beetles. Carpet beetles thrive on soiled clothes. Always clean clothes before storing them.

■ **Locate the hot spots.** Many times you will see adult beetles wandering about, but the main infestation will be localized in one or more spots. It takes some thorough detective work to find these hot spots. Check in particular the back sides of area rugs, stored clothes and blankets, places where lint and debris accumulate, and bags of pet food.

Carpenter ants (and other species of large ants) have been used for centuries

■ **Throw away or dry clean infested items.** Check all clothes and stored woolens. Badly infested and damaged items should be thrown away. Dry cleaning will kill all stages of carpet beetles. Infested woolens can also be brushed with a stiff brush to destroy larvae. Check pockets, cuffs and seams carefully.

■ **Vacuum fanatically.** Make sure to concentrate on those areas you usually skip like under baseboard heaters, along edges of wall-to-wall carpeting, and the back sides of rugs and furniture. When vacuuming to control bugs, you should throw out the vacuum cleaner bag after each use, or you can place a piece of flea collar into the bag. Try to reduce clutter.

■ **Protect valuables in sealable plastic bags.** Items that are known to be free from carpet beetles can be stored in sealed plastic bags. Camphor in tightly closed containers is said to kill carpet beetles.

■ **Keep windows screened.** This may help prevent adult beetles (and other creatures) from entering the house. Another possible source of infestation is cut flowers brought into the house. You can't worry about everything though.

■ **Use insecticides as a last resort.** Because carpet beetles are often localized, it doesn't make a lot of sense to treat your whole house with insecticides. Once you find the hot spots, and all else has failed, a pyrethrin-based insecticide (such as **XClude**) can be applied directly to the infested areas. Repeat the treatment in two weeks if larvae are still present.

■ **Don't give up.** It may take you a while to figure out what is going on and to get the upper hand. Remember what Calvin Coolidge said, "More often than not, persistance, rather than genius, leads to success."

to suture wounds closed. The ant is induced to bite the wound shut, and then

CLOTHES MOTHS
(Tineidae Family)

O robe, thou art sick!
The invisible worm
That creeps in the night,
In the closet warm,
Has found out thy bed
Of blue plaid joy:
And his dark secret love
Does thy life destroy.
(Apologies to William Blake)

I used to think that clothes moths were a pest that only got the other guy until while researching this book, I noticed that my favorite though seldom worn wool robe, had been altered into a garment more closely resembling lingerie.

My wife, being a wool rug maker, has always cast a covetous eye on moth-damaged wool items. Though she will not get my robe, she charitably offers to accept any unwearable, but braidable wool items readers care to send her way.

DESCRIPTION

There are several species of clothes moths (or wool moths) found in homes. The most common is the webbing clothes moth *(Tineola bisselliella)*. Adult moths are a golden or buff color with a distinctive tuft of reddish hairs on their heads. Larvae are about 1/2 inch long and a creamy white color when full grown.

Clothes moths are often confused with Indian meal moths, which are actually much more common in homes. Indian meal

its body is cut off, leaving the head in place. ♦ Bee swarms, which can contain

moths have bi-colored wings (half gray, half copper) that easily distinguish them from clothes moths.

LIFE CYCLE

Female moths, who remarkably never eat during their two week life span, lay eggs on woolen clothes, furniture, carpets or stored wool. The larvae hatch out and usually feed for about a month, but can feed for up to two and a half years! They pupate in a silken pupal case in a protected place attached to the fabric they were feeding on.

RECOMMENDED CONTROL MEASURES

Control measures for clothes moths are much the same as for carpet beetles. This is serendipitous as the two pests follow each other alphabetically in this book. Turn to page 18 for carpet beetle recommendations.

thousands of individuals, are completely harmless. Bees in general only sting

22

COCKROACHES
(Blatella germanica and others)

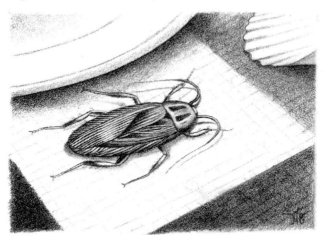

A True Story: A woman found a roach in her bathroom, threw it in the toilet, and sprayed it with a whole can of hair spray. Her husband came home, threw a cigarette in the toilet starting a fire that burned his private parts. The paramedics hearing the cause of the injury laughed so hard that they dropped him down a flight of stairs breaking several bones.

DESCRIPTION

The German cockroach is the most common cockroach found where I live in the Northwest and in most parts of the country. Fortunately it is too cool here for most other species to become established. Actually the Northwest climate is not especially suited to German roaches either, but if they are well fed they manage to survive.

In warmer parts of the country, many other cockroach species

when threatened. ♦ Collecting beetles was a highly respected hobby of 19th may be found including the brown banded, American, brown, oriental, Australian, and Surinam cockroaches, to name a few... (The common names of cockroaches vary from country to country and have little to do with the actual country of origin as much as the *imagined* country of origin. What might be called a Russian cockroach in Germany might be called an American cockroach in the Soviet Union or a German Cockroach in our country.) Have your extension agent identify your species and give you tips on its life cycle and habits.

Adult German cockroaches are about 1/2 inch long, pale brown (blackish when dead), with two parallel black streaks on the back of the head. Nymphs vary in size from 1/8 inch when they first hatch out, up to 1/2 inch. They are dark at first but become lighter as they grow. Eggs are laid within egg cases 1/4 inch long that contain thirty to forty eggs and look like little brown coin purses.

LIFE CYCLE

German cockroaches are found wherever they can find food, water, warmth and shelter. They are most commonly found within kitchens and bathrooms, but can be found anywhere that their needs are met. Roaches are attracted to dark cracks and will congregate in these areas. Areas such as wall voids, under cabinets, in motor housings, and under loose baseboards are all potential hiding places.

Adult females produce four to eight egg cases within their lifetimes. Each egg case is carried by the female for thirty days until just before the nymphs are ready to emerge. Nymphs molt six or seven times before becoming adults. There are up to four generations per year.

Roaches are nocturnal insects and are rarely seen during the

century English clergymen. ♦ To determine temperature accurately without a

day unless their numbers are very large. They will eat anything that vaguely resembles food.

RECOMMENDED CONTROL MEASURES

■ **Sanitation.** While miracles do happen, no one gets rid of roaches without cleaning up. While completely eliminating sources of food and water is often impossible, the cleaner your home, the easier it will be to get rid of the roaches. Never leave food out overnight. Empty trash containers every evening. A thorough 'spring cleaning', concentrating on areas that are not regularly cleaned, such as under refrigerators and stoves, is always helpful. In single family homes in the Northwest that are kept squeaky clean, roaches are fairly unusual. Unfortunately this is not so in warmer parts of the country.

■ **Caulking.** Roaches are able to move about in cracks as small as 1/16 inch, and feel right at home in cracks just 3/16 inch tall. Not very big. Caulking eliminates these roach cracks. Caulking efforts should be centered in the kitchen and bathroom. Any caulking labeled for indoor use can be used.

For various, mostly indefensible reasons, many people resist the idea of caulking. The major objection – "I will just make a big mess" – can largely be countered by using clear caulking which hardly shows, and a very small tip on your caulking gun. Caulking is always worth the effort.

■ **Sticky traps.** Used one per room, sticky traps help you determine where your roach problem is centered so that you can concentrate your attack where it will count. Used together in groups, they can catch enough roaches to put a significant dent in a population. To 'trap out' a roach population, use a minimum of four traps in each problem area, locating the traps against walls.

---

thermometer, use a cricket. Count the number of chirps in fifteen seconds and

■ **Boric acid dust.** For serious infestations, boric acid dust can be applied very lightly in baseboard cracks, wall voids, false ceilings, under refrigerators... every place a cockroach might hide. Boric acid dust has been used for generations and still works great. While of fairly low toxicity, never apply boric acid dust in areas where children or pets could get into it.

Boric acid dust is most easily applied from an aerosol can (such as **Perma-dust**), but can be applied with a hand duster or even a teaspoon when bought in bulk (i.e. **Roach-prufe**). Only a very small amount is needed to be effective. In fact, the lighter the dusting, the more effective it will be. A second or two from an aerosol can is sufficient for most small areas. Aerosols do contain unfriendly solvents which may be a reason to avoid using them.

Plastic squeeze-type ketchup or mustard containers can be used as dusters. Apply the dust as lightly as possible. Fill the container only half full for best results.

Boric acid can also be applied as a paste bait (i.e. **Blue Diamond Magnetic Roach Bait**). Small dabs of paste can be placed in infested areas in locations out of sight of kids.

■ **Cockroach growth regulators.** Especially in areas where cockroach levels are high or chronic, growth regulators can be very useful. Like their counterparts, the flea growth regulators, they are in effect, birth control for bugs. Growth regulators are of low toxicity and last for about four months.

When first applied, the solvents in the growth regulator may have an unpleasant smell that usually is gone within several hours. Wear a respirator and have windows open when applying. Roach growth regulators (i.e. **Torus** and **Gencor**) can be hard to find not pre-mixed with other chemicals, but are available. If necessary, a pest control company can apply these for you.

---

add 38. ♦ Foxes have been observed to use this ingenious way of ridding

FLEAS
(Ctenocephalides spp., Pulex irritans)

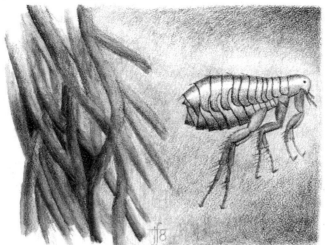

An airborne flea magnified 12x actual size

Have you ever returned home from vacation and been attacked by a house full of fleas? This is a fairly common experience, especially among pet owners. In parts of India they solve this problem by hiring a poor person off the streets to enter the house before the returning family, so the fleas will bite him instead of them.

DESCRIPTION

Fleas are about 1/8 inch long, dark colored, wingless, and exceedingly narrow-bodied. They have very strong legs that enable them to jump incredible distances. They also have a nasty bite. On most people it causes a small, red, itchy bump that goes away in a couple days. But some people have an allergic reaction that causes flea bites to swell painfully.

themselves of fleas. First they collect bits of hair and wool from bushes and

LIFE CYCLE

The three species of fleas which generally attack people are the cat flea, the dog flea, and the human flea. Unfortunately, fleas are not very good about staying on the host for which they were named. As most people with pets know, a cat flea will quickly become a human flea and is in fact, the most common flea on dogs too. The life cycles of all three fleas are fairly similar.

A female will lay four to eight eggs after each blood meal. (This means *you.*) Eggs are deposited near the site of feeding, not on the animal. The eggs hatch in about ten days and the new larvae feed on dried blood that the female leaves around for them. Depending on the weather and the time of year, it can take from one week to several months for the larvae to pupate, and another week or up to a year for the pupae to become adults. Adults can live up to a year under optimum conditions but more commonly live about two months.

RECOMMENDED CONTROL MEASURES

A serious flea control program is not for the weak or meek as it requires a serious, on-going effort to get the desired results. Try to incorporate as many as possible of the following suggestions into your plan of attack.

■ **Vacuuming.** Thorough vacuuming on a regular basis is essential to bring down flea numbers and remove sources of food. The vacuum cleaner bag should be emptied often (away from the house), or a piece of flea collar can be placed within the vacuum cleaner bag so a reservoir of fleas will not build up.

■ **Steam cleaning.** True truck-mounted steam cleaning (as opposed to the 'steam cleaners' you can rent at supermarkets) will quickly lower flea numbers in a flea emergency. Rarely will this get

fences and gather them into a hairball which they place in their mouth. Then they

28

the last flea, but it will always help.

■ **Flea comb your pets.** A flea comb is a specialized tool for removing fleas from short-haired animals. (It will also work on the chests and bellies of longer haired pets.) A flea comb is a small comb with metal tines spaced closely so that as the animal is combed the fleas get caught in the tines. The fleas can then be flicked out of the tines with a fingernail into a container of soapy water. If you have a pet, but not a flea comb, drop this book and get one now!

■ **Citrus-based Flea Dips.** For pets that can't be flea combed (fur too long or thick, or too ornery), other strategies have to be taken. There are several citrus-based flea dips of very low toxicity that do a good job of killing fleas. One brand that seems to work well is **Flea-Stop**. Some dogs reportedly enjoy being vacuumed. Sounds like fun to me.

■ **Restrict your pet to certain areas.** Limiting the whereabouts of your pets helps limit where fleas get spread. If possible try to make your pet exclusively either an indoor or an outdoor pet. If this is not possible, try to confine him or her to certain parts of the house, preferably areas without carpets.

■ **Have the pet sleep on a towel.** Most flea eggs are deposited on the ground where the pet sleeps. By having your pet sleep on a towel, most eggs will end up on the towel, and not on your carpet. Wash the towel twice a week so that the eggs do not have time to hatch.

■ **Flea growth regulators.** Flea growth regulators can be sprayed onto carpets to prevent juvenile fleas from reaching adulthood. It may sound like a bad thing to have a mess of pre-pubescent teen-age fleas in the house, but this is really a good thing. Teen-aged fleas neither bite nor reproduce, thus breaking the flea's life cycle.

---

slowly enter a river – backwards – until only their nose and mouth are not

Flea growth regulators remain active for four months. Two brands currently available are **Precor** and **Torus**.

■ **Linalool.** Not an exotic ethnic food, linalool is a derivative of citrus peels that can be used for flea problems. For heavy infestations that need immediate attention I have had good results with linalool containing a product called **Demize**. It can be applied with a clean garden sprayer, or a professional can apply it for you.

■ **Safer's Insecticidal Soap.** With something approaching cult status among organic gardeners, **Safer's Soap** can also be used in a two percent solution against fleas. It can be applied to carpets and floors with a clean garden sprayer. It remains active only while it is wet. While 100% control cannot be expected with a single application, repeated treatments every few days, as needed, will help.

■ **Diatomaceous Earth.** D.E. is a mined product that can be applied very lightly to animals (but not humans) and to carpets. The dust both smothers and dries out the fleas and their eggs. An ounce or two per dog should work.

Important: Never use the heat treated diatomaceous earth used in swimming pool filters as this causes the disease silicosis. Use only the natural mined product sold as an animal supplement or for garden use. Wear a dust mask when applying the dust. Don't use this method if you or your family spend a lot of time on the carpet.

■ **Keep your lawn mowed short.** And keep the lawn on the dry side. This will help lower outside flea populations.

The following is a simple flea trap that works best when pets are not around. You think it's too simple to work? Try it and you'll be surprised.

submerged. Then they drop the hairball and leap to land, leaving the fleas behind

The warmth of a low watt light bulb hung 6"–12" above an open, flat insect sticky trap will attract fleas, especially if other heat sources (i.e. animals) are not present. While it is unlikely that this trap will completely control a serious flea problem, it does make a good demonstration and can be used for monitoring purposes.

on the hairball. (Similar to the fox who "helps" the gingerbread man cross a river

FRUIT FLIES
(Drosophila spp.)

Time flies like an arrow but fruit flies like a cherry (...or a banana)

Most of the early genetic research done in this country was done using fruit flies. They were chosen because of their short life cycle (you probably knew that from personal experience!) and readily observable physical characteristics. It is hard sometimes to think of insects as useful, but we literally owe our lives to them. Without insects in the food chain, and as pollinators, there would be no life on this planet, at least not as we know it. It is undoubtedly part of Nature's master plan that the really essential things needed for life (air, water, sunlight, soil, insects...) are found in such great abundance. Unfortunately, sometimes there can be too much of a good thing... which brings us back to fruit flies.

coaxing him closer to his nose and then eats him.) ♦ It was rat fleas who were

DESCRIPTION

Adults are 1/8 to 1/4 inch long, dull, brown colored flies with distinctive red eyes. They are small! The larvae are even smaller with a strange-looking breathing tube sticking out of the tail-end. Pupae are brown and look like seeds.

LIFE CYCLE

Fruit flies are found on ripe and fermenting fruits, vegetables, vinegar, wine, and areas associated with any of the above. Actually, fruit flies are misnamed in that they are really attracted to the yeast associated with fermenting foods. They are most common in the fall when rotting fruit is most abundant outside (and inside), and decline in numbers quickly as the weather turns cold. Eggs are laid on any of the above food items and hatch within 24 hours. Larvae develop in five to six days. A complete life cycle takes 10 days. Each female can lay up to 500 eggs.

RECOMMENDED CONTROL MEASURES

Fruit flies are not a major threat to mankind but can be annoying. Always try to identify and remove the food source that is attracting the flies. Keep ripe fruit in the refrigerator. Two simple homemade traps often work well if competing food sources are removed. Fruit flies can also be kept away from a food source by blowing a small fan on the food so that the weak flying fruit flies can not land.

It is never necessary to spray for fruit flies. If all else fails, just bide your time for a few weeks. They are usually seasonal and go away by themselves after a while.

---

the vector that spread the bubonic plague that over several epidemics killed

TWO HOMEMADE FRUIT FLY TRAPS

nylon mesh screen

1/2 inch opening at
bottom of cone

1/2 vinegar, 1/2 water,
dash of dishwashing
soap

piece of fruit or
1/2 inch of vinegar

Replace fruit every two days or you will start breeding new fruit flies.

more than 100 million people (not to mention a lot of rats). ♦ Many butterflies

HEAD LICE
(Pediculus humanus capitus)

A friend reported to me how worried his child was when the nurse told him his hair was covered with head lights. You would be too.

DESCRIPTION

Head lice are gray in color and just under 1/4 inch long. They are shaped something like ticks, only smaller. Their eggs, called nits, are attached to hair strands with a powerful glue. In practice, it is the location of the head lice, the accompanying itch, and a school memo, that identifies them.

A head louse emerging from a nit on a strand of hair

LIFE CYCLE

An adult female will lay up to 150 eggs that hatch in five to ten days. The eggs (nits) are glued directly to hair right at the scalp line. The nymphs feed on blood, molt three times, and complete their life cycle in about three weeks. Adult lice live about three weeks too.

RECOMMENDED CONTROL MEASURES

There is much misinformation associated with head lice. Some people think that only dirty people get head lice. This is not true.

use bright patterns to advertise that they are foul tasting. Other good tasting

Others are convinced you need to use pesticides to be free of head lice. This is not true either.

The following is the non-chemical procedure for treating head lice. It works!

■ **Hands off!** Head lice spread with incredible ease. Even brief physical contact can spread them. At least for the duration of the infestation, avoid all physical contact with infested people. Parents should avoid hugging their kids. Kids need to be told to fight from a distance.

■ **Don't share personal items.** Head lice are very easily spread by combs, towels, and shared clothing. You are officially sanctioned not to share anything with anyone during the treatment period.

■ **Use a head lice comb and a coconut-based shampoo.** A head lice comb is a specially made comb with closely spaced metal or plastic tines that is used to strip off head lice and nits.

To use a head lice comb properly, follow these instructions:

1. Wash head with shampoo and do not rinse.

2. With head wet, methodically comb inch wide clumps of hair starting as close to the scalp as possible.

3. Keep the head wet the whole time to prevent unwanted hair removal.

4. Rinse shampoo out of hair when finished. Re-check hair for any missed nits.

5. Repeat weekly for the duration of the problem until the last person at home or school is free of head lice.

This procedure takes time to do right, but is a good way to

butterflies will copy their bright colors and patterns to appear as if they are

spend some 'quality time' with your child or spouse, and, you avoid using pesticides. If your school nurse is unfamiliar with this method or you would like more detailed instructions, send for the pamphlet 'IPM for Head Lice' available from: Bio-Integral Resource Center, P.O. Box 7414, Berkelely, CA 94707 ($4).

■   **Never spray a house for head lice.** Head lice can live only a short while away from a nice warm human body – a few days at most, usually less. So there is never a need to spray your home. Remember also that there is no need to use the various insecticidal shampoos and ointments if you are willing to take the time to use a head lice comb carefully.

HOUSEFLIES
(Musca domestica and others)

One time I was called to a doctor's office that was having an on-going problem with flies. They were located on the top floor of a new medical building with no opening windows. I puzzled over how the flies were getting in. After much detective work, I figured out the flies were entering the front door of the building, flying up the elevator shaft into the false ceiling, and entering the office through little ventilation holes in the florescent light fixtures.

of holiness. The more lice, the holier the man. ♦ Blowfly larvae maggots were used

DESCRIPTION

Most people are not aware (and probably don't care) how many different types of flies get into homes. There are lots. The true Domestic Housefly adult is 3/16 to 5/16 inch long, and has four dark stripes on its back, two wings and large red-brown eyes surrounded by a light gold stripe.

LIFE CYCLE

Female houseflies lay their eggs in moist organic matter including manure, garbage and compost. The eggs are laid one at a time but in clusters of up to 150. A female may lay more than 500 eggs in her short lifetime.

The eggs hatch in a single day and the larvae burrow into their food source and feed for up to several weeks. The larvae move to a drier location before pupating which can take from three days to four weeks. Under optimum conditions (for the fly) their life cycle can be completed in one week.

RECOMMENDED CONTROL MEASURES

■ **Exclusion.** Without a doubt the key to good fly control is exclusion, that is, keeping them out. I know this isn't earth shaking, but screens on doors and windows are what it takes to keep flies out of homes. The kids have to be trained to keep doors closed. The adults have to be reminded to make sure the screens are in place and in good shape.

■ **Keep your home clean.** Flies are constantly on the lookout for food and breeding sites. Having food lying around is like putting out a welcome mat and a red light for flies. Ideally, garbage should be removed twice a week so that fly eggs don't have enough time to hatch.

in the Civil War and afterwards to clean out wounds, the maggots limiting their

■ **Fly swatters.** Maybe these aren't 'high tech' enough for you, but, especially when fly numbers are low, fly swatters are very effective, and fun! Flies are easiest to swat earlier in the morning when temperatures are cooler. Flies will always fly away straight up and slightly backwards. Remember, it's all in the wrist...

■ **Fly paper.** Another old fashioned remedy is fly paper. Recently, new types of fly papers and fly traps have hit the market. Some contain food or pheromone (insect hormone) attractants. Others rely on shiny reflective surfaces to attract flies. These can be useful in kitchens as long as they are not placed directly over food or food preparation surfaces. But you've still got to have screens on the windows...

■ **Vacuum cleaners.** Sometimes when you have a swarm of hovering, circling male flies you can suck them from the air with a vacuum. This can be lots of fun. 'Dustbusters' are not strong enough. A powerful vacuum with a long extension tube works best.

■ **Don't use sprays or 'pest strips'.** Neither deals with the source of the problem and both pollute your house. Screen doors are a better investment.

■ **Fly traps and fly parasites.** Outside, there are several types of fly traps that can be used. In stables, commercially available fly parasites can be applied to fresh manure to destroy eggs before they hatch.

I asked an artist if he could do insect illustrations. He replied, "I don't see why not. All summer long I draw flies."

---

diet to decaying flesh. ♦ The housefly beats its wings to the pitch of an F in the

INDIAN MEAL MOTHS
(Plodia interpunctella)

M y wife and I have this (usually) friendly game where she swats the meal moths in the house and I try to rescue them. Most of the time I enjoy them quietly fluttering around. I see them as indoor butterflies. She sees vermin.

DESCRIPTION

Indian meal moths are one of the more common pests of stored products and are the most common small moth in homes. The adults are less than 1/2 inch long with copper colored wings with a broad gray band where they attach to the body. They are weak fliers. Indian meal moths are often confused with clothes moths, which are slightly smaller in size and have buff colored brown and a reddish tuft on their heads.

middle octave at 345 strokes per second. ♦ "If a man make' a better mousetrap

The larvae are small caterpillars that vary in size from 3/8 inch to 3/4 inch long. They vary in color from off-white to pink, brown, or light green. The head and first body segments are brown.

LIFE CYCLE

The larvae are found on a variety of stored goods. They have eclectic tastes in food preferring coarse grains, such as bran and whole grain cereals, but also infesting such diverse food items as dried fruit, nuts, beans, chocolate, dog food, and dried chili peppers. Within the infested food, they produce a webby material that contains the insect frass (feces).

Indian meal moths usually overwinter as larvae. Adult females lay their eggs at night on the larval food, producing up to 400 in 18 days. The larvae live and feed within their food source, sticking close to the webby material they produce. They pupate in cracks away from the webby mat.

There are four to six generations a year depending largely on room temperatures and the food source.

RECOMMENDED CONTROL MEASURES

If adult moths are seen flying around, a thorough search needs to be made for the infested food source from which the adults emerged. The infested food source can be readily identified by the presence of the webby material and the larvae themselves. This food should be disposed of immediately. If the storage container is to be re-used, it should be washed thoroughly.

Pheromone traps are very effective in controlling meal moths. These are special traps that contain lures with insect pheromones (hormones) to attract male meal moths and a sticky glue that they get stuck in. (What a way to go...) With the males removed from

than his neighbor, tho' he build his house in the woods, the world will make a

the population, no mating takes place, breaking the life cycle.

Traps should be placed at the rate of one trap per every 1000 cubic feet (or 120 square feet of floor space except where ceilings are high) in areas where adult moths are commonly seen. Usually one trap in the kitchen is all it takes. The lure in the trap needs to be replaced every three months. **Surefire** is one of the more commonly available brands.

Pheromone traps are sold with different lures to attract different pests. But just like when fishing, you've got to use the right bait. Make sure you get the right lure.

---

beaten path to his door." – Ralph Waldo Emerson. ♦ A single silkworm cocoon

MICE AND RATS
(Mus musculus, Rattus rattus, Rattus norvegicus)

HOUSE MOUSE: 6" – 7" long; grey color; prominent ears.

A fter trapping mice all day, it is a little strange to see my daughter at home playing with her pet mice, Angelina and Carrie Anne. It just goes to show how little difference there is between a pet and a pest.

DESCRIPTION

Two kinds of rats and a single mouse species are sometimes found around homes. These are the Norway rat, the roof rat, and the common house mouse. The Norway rat (or brown rat) is the most common species of rat in both the Northwest and the United States. The roof rat or (black rat) is found mostly near coastal areas, and very rarely east of the Cascades. Mice are found everywhere.

There are several characteristics which can be used in telling them apart. Notice in the illustration the different relative sizes of the various body parts.

The different rats and mice can be also distinguished by their fecal pellets or excrement. (This may sound disgusting, but is very useful when only their 'calling cards' are present.) Fecal pellets of the Norway rat are 3/4 inch long and have blunt ends. The fecal

produces half a mile of silk fiber. 3,000 cocoons produce one pound of silk. ♦ "A

pellets of roof rats are 1/2 inch long and have pointed tips. The house mouse fecal pellets also are pointed but are only 1/4 inch long.

The presence of rats and mice can also be determined by looking for urine stains in suspected mouse infested areas with the aid of an ultraviolet light (a 'black' light). Rodent stains will 'glow' if they are present. Rodent urine also has a strong, distinctive odor.

NORWAY RAT: 12" – 18" long; red brown/grey color; smaller ears; thick set; blunt nose.

LIFE CYCLE

Mice reproduce with amazing speed. At 35 days of age a female mouse is mature. In 21 more days she can give birth to her own kids. Litter size is usually six. Considering the offspring of her offspring, that means almost 8,000 offspring per year from one mouse mama! Young are nursed for four weeks. Indoors, mice can breed year-round. Their life expectancy is less than one year.

The Norway rat reaches maturity at two months; the roof rat at four months. Average litter size is seven. There may be three to six litters each year. Average life expectancy is about six months. Most rodents are primarily nocturnal.

RECOMMENDED CONTROL MEASURES

Mice and rats can be controlled in most cases through a combination of trapping and excluding them from buildings.

man thinks/ he amounts to a lot/ but to a mosquito/ a man is/ merely/ some-

■ **Snap Traps.** Having withstood the test of time, snap traps should be your first recourse for the occasional mouse in the house. Even large numbers of mice can be caught quickly by placing out a bunch of traps at one time. For the coddled homeowner, I recommend an improved, easy-to-set mouse and rat trap called the **Snap-E** trap. It is made so that you never touch the dead mouse.

Snap traps need to be placed carefully to be effective. Whenever possible they should be used in pairs and placed along walls, end to end, with the triggers facing out. A well positioned trap often does not need bait, but a small amount of peanut butter or Reese's peanut butter cup can't hurt.

Here's a hot tip for catching a a wily mouse or rat. Try placing the traps out for several days baited, but not set. Once the creature starts eating the bait, then the traps can be set, usually with good results. Rats are especially bait shy, and more patience is needed in trapping them.

ROOF RAT: 13" – 17" long; black/grey color; larger ears; slender build; pointed nose.

■ **Multiple Catch Traps.** For mice, the multiple-catch traps such as the wind-up **Ketch-all** and the newer **Victor Tin Cat** traps can catch ten or more mice in a single setting. These traps are best suited for garages and other places where a trap full of mice won't alarm people. It should be placed parallel to and an inch or so from the wall, in areas that mice are seen in.

thing to eat." – Don Marquis. ♦ While spending a summer in the Forest Service

As these traps are 'live traps' they must be checked daily to see if any mice have been caught. Mice can either be killed or released. Humans are usually quite resourceful when it comes to killing things, but for those of you with less imagination, the simplest method is probably to flush them down a toilet. Less hardy souls can release the mice in fields to give them a chance to come back again. More sporting.

■ **Mouse-proof your house.** While trapping can temporarily clear up a mouse or rat problem, unless you determine how the creatures got in the house, and do something about it, others will sooner or later follow. Somewhere you will find a crack or hole in a wall, or an enlarged pipe entry, an unscreened crawl space vent, or a door with a big gap underneath, where the creatures are entering. These should be sealed. Steel wool makes a good temporary seal for many holes. Weatherstripping should be used under doors. Mice need only a 1/2 inch gap to squeeze through, so entry holes need to be looked for carefully.

■ **Poison Baits?** Some people like to use poisoned baits to control mice. I do not recommend these for several reasons. First, you are dealing unnecessarily with poisons. Second, you run the risk of a pet or other animal being poisoned from eating a poisoned mouse. And third, sometimes mice will die within walls creating a foul odor.

If you still want to use baits, look for one with Vitamin D–3 *(cholecalciferol)* as the active ingredient. This product is still poisonous to people and pets, despite containing a vitamin as the toxicant, but is probably least hazardous, all things considered.

■ **An ounce of prevention is worth a pound of cursing.** Mice are most likely to enter homes as the weather turns cool in the fall, so try to tighten up the house before you have a house full of mice.

I lived for a while in a cabin infested with mice. Aside from the smell, the only

SILVERFISH
(Lepisma saccharina)

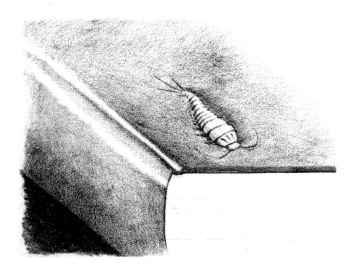

Occasionally while inspecting under a house I find an old newspaper, as much as 50 years old. Sometimes it will be partially eaten by silverfish. The interesting thing to me is how *little* damage was done in all those years. While probably capable of doing serious damage to paper products, I feel silverfish have to be near the top of the list of over-rated pests.

DESCRIPTION

Silverfish are wingless insects with slender, scale-covered carrot-shaped bodies, two long antenna in front, and three bristles in back. When fully grown they are about 1/2 inch long. They scurry around quickly when disturbed, and seem to upset people dispro-portionately to their size and destructive potential.

serious consequence of living with mice was they ate a significant amount of my

LIFE CYCLE

Silverfish are found in dark, damp, and often warm locations within buildings. They feed primarily on paper, the glues found in the sizings of books, and cereals. Occasionally they feed on dead insects or dried meat. Females lay up to three eggs at a time and lay about 120 eggs in their lifetime. The eggs hatch in about six weeks under optimum conditions. It takes three or four months for young silverfish to reach adulthood. Silverfish often live up to three years, quite a long time for an insect.

RECOMMENDED CONTROL MEASURES

■ **Stop and think.** The first thing to do is to look deep into your soul and determine why you want so badly to kill these basically harmless creatures. They are found in small (or large) numbers in most buildings everywhere and, for the most part, do very little damage. (A few exceptions – stored books and papers, and art supplies.) Granted they are uninvited, but so are my visiting relatives. In comparison, silverfish are better behaved, quieter, and eat much less. (Just kidding, Mom...)

■ **Boric Acid Dust.** Nevertheless, silverfish can be controlled by placing boric acid dust in the cracks, and other areas where they live (such as attics or garages). Where silverfish are found in non-storage areas of buildings, boric acid dust can be applied under loose baseboards and in the spaces *around* electrical outlet boxes (be careful working near electricity!) The baseboards can then be caulked if the problem is serious. Attics and other infrequently used storage areas can be covered with a light dusting of boric acid dust.

Boric acid dust is available in three forms: in bulk, in aerosols, and as a paste bait (used mostly for roaches). In bulk (i.e. **Roach-**

---

book, The Little Flowers of St. Francis Assisi, patron saint of animals. ♦ "Happy

Prufe) it can be applied from a plastic squeeze bottle such as a mustard or ketchup dispenser. Fill the container only half way so it can be squeezed.

The aerosol form (such as **Perma-Dust**) is somewhat easier to apply but contains solvents which smell bad and can't be good for you. In either form, apply the dust as lightly as possible; the lighter, the better.

■ **Control Humidity.** Silverfish like it damp. If possible, try to reduce the humidity in areas where silverfish are found. Fixing plumbing leaks and providing adequate ventilation will sometimes help reduce their numbers. Getting rid of old boxes full of junk in storage areas may also help.

One thing to keep in mind – it is simply impossible to get rid of every last silverfish in your home. Might as well get used to having a few around for company.

---

the cicadas' lives, for they have voiceless wives." – Xenarchus, a male chauvinist

SPIDERS
(Many species)

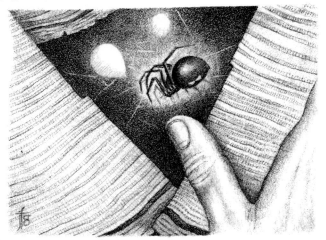

One summer, a large, beautifully patterned, orb weaver spider took up residence in my bedroom window. It did fine for a long time catching flies, which made me happy. In the fall, when houseflies became scarce I brought fruit flies from the kitchen to its window. It was a sad winter day when cold weather finally killed it. I felt like I had lost a friend.

DESCRIPTION

There are many species of spiders that sometimes get into the house. Most people break them down into three categories: black widow spiders, brown recluse spiders, and all the rest. Fortunately, where I live in the Northwest the black widow is not all that common, especially west of the Cascades. The brown recluse is also a very rare visitor. The other spiders are fairly benign with *one* exception, the aggressive house spider, which can produce a *nasty* ulcerous wound.

ancient Greek. ♦ "The ant has made himself illustrious,/ through constant

- **Black Widow Spider** females, the ones who bite, are about 1/2 inch long and shiny black with a red hourglass shape on the underside of their *plump* abdomens. Black widows are fairly common in warm parts of the country.

- The **Brown Recluse Spider** is up to a 1/2 inch long, comes in various shades of brown, and has a violin shaped marking on the underside of its middle section (thorax). It is most common in Oklahoma, Missouri and surrounding states.

- The **Aggressive House Spider** *(tegenaria agrestis)* is a common spider within homes. It is brown with chevron (v-shaped) patterns on its abdomen. It is a swift running spider, is over an inch long, and makes funnel shaped webs. (Other funnel web spiders are also common in homes but are non-aggressive and do not cause painful bite reactions. Most do not have the chevron pattern.) House spiders are found throughout the country.

LIFE CYCLE

The different species of spiders have somewhat similar life cycles, but because there are so many species, it is hard to generalize. Most lay 200 or more eggs in egg sacs. Eggs hatch in about a month, and the little spiderlings pass through several instars (stages) before reaching adulthood. The complete life cycle usually takes several months.

RECOMMENDED CONTROL MEASURES

■ **Calm down.** Without question, the first thing to do is to relax. Odds are you are not dealing with a very poisonous or even a slightly poisonous spider. Try to capture the spider that bit you so it can be identified. Even the squished remains can often be identified. See a doctor if you are experiencing abdominal or

industry industrious./ So what?/ Would you be calm and placid,/ if you were full

extreme pain. Make sure you take your spider with you, if possible.

■ **Sweep them up.** If the offending spiders *are* in the house, take a broom and sweep (or vacuum) every corner of the house including corners of walls, ceilings, bookshelves – everywhere! This effectively destroys spider homes and squashes some spiders. If this is repeated weekly for several weeks, any surviving spiders will not have time to rebuild nests and lay eggs, thus breaking their life cycle.

If you suspect you are being bitten in bed, and the culprit is not your spouse or child, wash your sheets and covers, or brush them thoroughly. Most spider bites in bed occur when the mattress is directly on the floor. If necessary, raise your bed off the floor.

■ **Hose them down.** Outside, if spiders are present in excessive numbers under eaves, simply knock them down with a strong blast of water with a hose. Again, repeat as necessary. Also keep grass and weeds cut low next to the house, and prevent shrubs from becoming overgrown.

■ **Wear gloves when working in spider habitat.** If you are worried about spiders, wear gloves while handling firewood, working under houses, or cleaning up junk around the yard. The two most serious spiders are not aggressive by nature. If you are careful not to threaten them, they will (usually) not attack you.

It is very important to try to remember that spiders are an essential part of nature's system of checks and balances. On farms, spiders play an important role in controlling other pests. Even in the house, spiders will help control houseflies and fruit flies. While our tolerance for all pests needs to be increased, spiders are, all things considered, such overwhelming good guys that we really need to ease up on them.

of formic acid?" – Ogden Nash. ♦ Lacewing larvae are beneficial insects in the

SUGAR ANTS
(Formicidae Family)

While going to college I lived in an ancient travel trailer on a farm that was under constant, all-out attack from Argentine ants, a type of sugar ant common in California. The only thing that slowed them down was when I painted the axles, and all pipes and wires leading to the trailer, with bands of **Tanglefoot**, a sticky non-drying anti-pest glue. This would deny the ants access to the trailer as long as no twig or blade of grass created a bridge over the glue. But as soon as *anything* touched the sticky bands, within hours, the ants were back inside in full force.

DESCRIPTION

There are several species of small ants, all commonly called 'sugar ants', that sometimes invade homes. In Oregon, my home turf,

garden that are voracious eaters of many kinds of insects and insect eggs. Their

and many other parts of the country, the most common ant found in homes is the odorous house ant *(Tapinoma sessile)*. This ant is small, 1/8 inch long, black or brown, has a spineless back, and a foul smell something like rancid butter when crushed. Thus the name.

In Washington, yellow ants *(Acanthomyops spp.)* and cornfield ants *(Lasius spp.)*, also commonly called 'moisture ants', are often found in homes. In California, the Argentine ant *(Iridomyrmex humilis)* predominates. Every area will have its most common species. Your local Cooperative Extension office can help you identify your species of ant.

(To collect any kind of insect without damaging it, gently flick it into a container with a tight lid and place it in the freezer. Insects given the 'cold treatment' for a few minutes will be slow enough to examine. Over night in the freezer kills most insects.)

LIFE CYCLE

New ant nests are usually formed when a portion of an existing nest splits off, forming a satellite nest. These nests can be found outdoors or indoors and are often located in wall voids. Moisture ants are usually found in or near moist wood.

Each nest contains several females which each lay a single egg a day. It takes from one to three months for an odorous house ant to reach adulthood. Other ants may have shorter or longer life cycles.

Odorous house ants usually show up in buildings during winter when 'honeydew' (a sweet excrement from aphids) and other food sources are gone. They prefer sweets, but will eat most anything.

---

own eggs are laid on the tips of slender stalks so as to escape predation from

RECOMMENDED CONTROL MEASURES

Sugar ants can be one of the more difficult pests to control. What they may lack in size, they make up for in sheer numbers and persistence. You too will need to show some persistence to gain the upper hand. The following methods when used together will enable you to successfully control sugar ants.

■ **Try to locate nest sites.** Controlling ants of all kinds is easy if their nests can be found. Finding a nest saves you hours of time in the long run and is always worth the effort. What you are looking for is not the nest itself, but the door to the nest – the crack the ants go in and out of. To locate a nest, simply follow a trail of ants back until they disappear under a baseboard or wherever...

If your ants are not walking in trails or are not numerous enough to follow, try this. In the evening, place out little cardboard or wax paper squares of pancake syrup or other sweet stuff along baseboards or the backs of counters where ants are likely to find them. In the morning, there should be good thick trails leading from the syrup to an ant crack in a wall somewhere.

■ **Keep your house squeaky clean.** Empty garbage daily. Keep the outside of jars wiped clean. Rinse out pop bottles and cans before recycling. Remember a very small amount of food can sustain an ant colony for a long time.

Cat and dog food dishes can be placed in 'moats' made out of saucers or bowls slightly bigger than the pet's dish and filled with water. Honey or sugar bowls can also be 'moated'. Ants as a rule don't like swimming.

■ **Caulk.** Set up a day to go around the infested area with a caulking gun sealing up all observable cracks and possible points of entry. This will help cut down on other insect pests as well. Very helpful! Keep your kitchen and bathroom dry. In many cases, ants

their own kind. ♦ Mice can jump down 10 feet without injury, leap up 12 inches, walk

63

are looking for water as much as for food. Keep counters wiped dry. Turn off water to toilet bowls. (Just kidding...)

■ **Ant Baits.** Some ants can be controlled using poison baits placed out in bait stations. Obviously these cannot be used in areas that are accessible to kids or pets. The most effective baits often contain a small amount of arsenic, which when used in the very small amounts recommended, can be used safely. Unfortunately (or fortunately), these products may be taken off the market soon by the Environmental Protection Agency (EPA).

Many bait stations come enclosed (and are therefore less hazardous to pets and children), and may contain boric acid or other insecticides. Boric acid works as a slow poison against the odorous house ant and may in the long run attract more ants than it kills. Despite often heard grumbling that baits don't work, baits *can* be used safely and effectively. You will have to experiment to find a brand that works for you.

Always write the date on any bait station you use. Bait stations have an active life about as long as a coconut custard pie. Discard *all* bait stations after a month.

■ **Pyrethrin insecticides.** Pyrethrin is a type of botanical insecticide derived from a species of chrysanthemum. It is often considered the safest of the household insecticides. (A few people are allergic to it.) A small amount of a pyrethrin insecticide (such as **XClude** or **Revenge**) when sprayed directly into a nest will easily destroy it.

■ **Look for outside nests.** Check around the outside of your home to see if ants are climbing up the foundation wall and sneaking inside. If you can find the outside nest, pour a pot full of boiling water directly onto the nest to destroy it. Repeat if necessary.

on telephone wires, swim (but don't like too), are color blind and have poor vision

If an outside nest cannot be located, a cross-shaped application of a micro-encapsulated pyrethrin-based aerosol (such as **XClude**) can be applied on to the foundation wall to keep the ants out of the house. The vertical line covers the ant trail itself. The horizontal line creates a barrier. The gesture of making a cross cannot hurt either. Go easy on the spray. A little often works as well as a lot. This can be repeated monthly or as needed.

■ **Talcum powder.** Believe it or not, baby powder will kill sugar ants, albeit slowly. Insects 'breathe' through their skin. Talcum, and other dusts, clog up their skin. Apply a very light dusting in areas where ants are known to frequent. Despite the name, baby powder should really not be used around infants as it is not good for people to breathe either.

■ **Sticky Barriers. Stikem Special** and **Tanglefoot** are two brand names of a sticky gooey material that can be spread in bands along foundation walls to keep ants out. These barriers never dry but do get coated with dust. This stuff is *very* messy and not beautiful to look at. It is probably only practical to apply to homes or parts of homes supported above ground on posts unless you have low standards of aesthetics.

---

but excellent sense of smell. ♦ In 1926 in the Central Valley of CA, the field mouse

TERMITES

(Zootermopsis angusticollis, Reticulitermes hesperus, Incisitermes minor)

Subterranean termites travel in "mud tubes" between their nests in the ground to the wood structure of your home

One of the surprising things about termites is how delicate they are. The workers especially are soft and squishy and until you see them in action, it is hard to imagine how they could possibly do any harm to your home.

DESCRIPTION

Fortunately for me, termites are not as big a problem in the Northwest as they are in warmer parts of the country. We commonly get two kinds of termites here – the dampwood termite and the subterranean termite. The drywood termite while common in California, is almost unheard of here.

Termites are often confused with ants. The main difference between the two is whether they have a 'waist'. Ants always have

population exploded to an estimated 82,000 mice per acre. ♦ "From red-bugs

three distinct body parts, head, thorax, and abdomen, with the thorax and abdomen separated by a narrow waist. Termites have only two body parts, a head, and a combined thorax/abdomen – no waist. Also, termites tend to be lighter colored.

• **The Dampwood Termite** is the most common termite in the Northwest undoubtedly because of its preference for damp or water-damaged wood of which there is no shortage here. Dampwood termites, like most termites, have three castes: reproductives, nymphs (who function as workers), and soldiers. Reproductives are light brown with inch long wings. They are often seen in weak flying swarms with the first warm rains of fall. Nymphs are cream colored with darker abdomens, and up to 1/2 inch in length. Soldiers have very large dark heads and are up to 3/4 of an inch long. They always live in or near damp wood.

• **Subterranean Termites** are a more formidable foe than dampwood termites. They live in the ground and travel through 'mud tubes' to the wooden structure of your house which they eat. Winged reproductives are black and about 1/2 inch long. Workers are white with darker heads and about 1/4 inch long. Soldiers have elongated heads and are slightly larger.

In some parts of the Deep South, the Formosan termite *(Coptotermes formosans)* has become established. This is a larger, more aggressive species that also builds 'mud tubes'.

• **Drywood Termites** are a common pest of the South and in California. Unlike subterranean termites, their nests may be found in any part of a structure including the attic. Drywood termites spend their entire lives in wood. Winged reproductives are dark brown and just under 1/2 inch. Larvae are white with yellow-brown heads. Soldiers have light colored bodies and massive red-brown heads.

---

and bed-bugs, from sand-flies and land-flies/ mosquitoes, gallinippers, and

LIFE CYCLE

Most termites raise large, well organized colonies with different tasks completed by different castes. Reproductives mate and lay eggs. Most of a colony are workers. Soldiers protect the colony from ants and other pests. A large colony may contain tens of thousands of individuals, an awesome sight!

RECOMMENDED CONTROL MEASURES

Termites are one of the few pests where the different species have to be treated differently.

DAMPWOOD TERMITES

In most cases controlling dampwood termites is simply a matter of eliminating the source of moisture, and removing and replacing the water damaged, infested wood. The reason this works is that dampwood termites must have ready access to moisture. Without it they simply cannot survive.

It is rarely ever necessary to spray for dampwood termites, the only exception might be in cases where eliminating the infested wood was impractical.

SUBTERRANEAN TERMITES

Several things make treating subterranean termites more difficult. For starters, they infest either wet or dry wood, so they may be found anywhere. Second their colonies can be hard to locate. The usual way to find them is to inspect under your house for 'mud tubes' connecting their nest to the wooden structure of your house. These mud tubes can be easily overlooked, especially if they are located behind concrete steps or in cracks in the foundation wall. Last but not least, subterranean termites are capable of doing the most serious damage the quickest of the various wood destroying insects, so the stakes are higher.

fleas,/ From hog-ticks and dog-ticks from hen-lice and men-lice,/ We pray, good

Since the 1940's, the standard way of treating 'subs' has been to create a chemical barrier between the termites and your home. For years **Chlordane** was the chemical of choice until very recently when all uses of it were banned in this country. Today other termiticides are used though none of them are without their drawbacks.

Recently several new treatment options have become available for subterranean termites. Only time will tell which will be really practical, but people who are not crazy about treating their homes with termiticides may wish to look into them. Somewhere among them, or nearby, is the future of termite control.

■ **Predatory Nematodes.** Nematodes are microscopic worm-like creatures that attack termites and many other kinds of insects. I have used these with some success. They are completely non-toxic, but provide no on-going protection (which is not necessarily a drawback). If you have a nest that can be monitored easily to see if the treatment was successful or not, give them a try. If the nest remains active after a month, try another treatment option.

■ **Sand Barriers.** Research has shown that a special sized coarse sand (twelve grit) when tamped down against foundation walls makes an effective termite barrier. To my knowledge this has not been tried in the Northwest but has been approved by Honolulu, Hawaii building codes and at least one company is trying it in California. The particular sand may be hard to find locally. Look for it from suppliers of sand blasting materials.

■ **Hot and Cold Treatments.** Several pest control companies use specialized equipment that alters temperatures either higher or lower than termites can tolerate. With the heat treatment, hot air is blown into the tented house until the termites are cooked. With the cold treatment, nests are identified and then spot treated with liquid Nitrogen.

Lord, give us ease." — 19th century prayer. ◆ A mouse in a month eats about five

■ **Borate Barriers.** Recently in Southern California, one pest control company has reported success using sodium borate as a chemical barrier in place of more toxic termiticides with good success. This would not be suitable for areas with high water tables as this chemical is water soluble. Also, despite its low toxicity, there is question of its legality as a termite control measure.

■ **Spot Treatments.** Many times a localized area can be treated instead of a whole house. Especially in the Northwest, infestations are often not widespread. A 'spot treatment' has the dual advantage of using less chemicals and costing less money. A thorough inspection is required to locate the 'hot spots' that require treating.

■ **Pyrethrin-based Termiticides.** If you have to go with a termiticide, the pyrethrin-based ones are considered less toxic and less hazardous to people and the environment. Some have less foul smelly solvents than others.

DRYWOOD TERMITES

A different approach has to be taken with drywood termites as they may nest in any part of the house. Traditionally tent fumigation has been used to control drywood termites as every part of the house is treated with a poisonous gas at the same time. But two less invasive methods – one old, one new – can be used instead:

■ **Spot treating.** Following a careful inspection to locate different areas of termite activity, each infestation can be 'spot treated' with a pyrethrin insecticide (i.e. **Tri-Dye**) directly into the nest.

■ **Electro-gun.** In some areas, pest control companies are using a patented device called an **Electro-gun** that uses a high voltage, low amperage current to treat drywood termites. This method is completely non-chemical and is worth checking into if it is available in your area.

---

ounces of food and produces about 1500 droppings. ♦ The reason the US owns

WASPS AND YELLOW JACKETS
(Vespula spp.)

One time I was called in to remove a huge baldfaced hornets nest 25 feet up in a tree in front of a school. Kids were being stung even though the nest itself was not in harms way. The school district provided me with a mechanized elevating platform that brought me up to the nest. To my surprise, the hornets paid no attention to me as I closed in on them, probably thinking themselves invulnerable so high up. I clipped the nest into a garbage bag, closed it up and headed down without a single sting.

DESCRIPTION

The names 'yellow jacket' and 'wasp' are used by different people to refer to different insects. Many people call anything that stings a 'bee' but this is really stretching things. After all, bees make

the Panama Canal and not the French who started it, is that the French gave

honey (usually). The following are the Northwest's most common 'social wasps', as the whole group is called. Other areas may have different species.

- **The Western Yellow Jacket** *(Vespula pensylvanica)* is the common, ground nesting yellow jacket that forages actively through the warmer months of the year near the ground.

- The **Aerial Yellow Jacket** *(Vespula arenaria)* looks like the western yellow jacket but makes large paper nests, usually under eaves of buildings.

- The **Baldfaced Hornet** *(Vespula maculata)* is a wasp with a white face and mostly black body that makes its beautiful paper nest in a tree or shrub usually several feet off the ground. It can be very aggressive.

- The **Umbrella Wasps** *(Polistes spp.)* look like long yellow jackets but make small nests consisting of a single comb without a paper cover around it. The nest can contain just a few cells or up to 250 (rarely). The umbrella wasps are very common under eaves and overhangs. They are not aggressive and should be left alone in most cases.

- The **Mud Daubers** (different species) make nests out of mud usually under eaves or in attics. They are also non-aggressive if left alone.

Many old-timers were happy to have yellow jackets make nests in their fruit orchards because they knew how important they were as predators of fruit damaging insects, such as the coddling moth. Even in urban areas, yellow jackets are beneficial insects in that they help keep down the populations of other pests, and should not be destroyed needlessly.

---

up after sustaining heavy losses from Yellow Fever spread by mosquitoes. ♦

LIFE CYCLE

Most social wasps have a similar life cycle. Only mated queens survive the winter, hiding out in sheltered places. In spring the queen establishes a colony laying up to 20 eggs. She builds a nest, lays eggs, and raises the first brood herself. Once the first workers are fully grown, the workers take over doing everything except egg-laying. Near the end of the summer new males and queens are produced. These fly away and mate. Cold weather kills the colony, including the original queen.

Adults feed mostly on sweet materials such as nectar and fruit juices. But because the larvae need protein, yellow jackets will hunt very aggressively for protein sources, including whatever you may be eating at your barbecue.

RECOMMENDED CONTROL MEASURES

Yellow jackets can be a serious problem, especially to people who are allergic to their bites. Fortunately, destroying nests in known locations is not as difficult as you might imagine, as long as you do it at night.

Unfortunately, nests cannot always be located, and even if a nearby nest is destroyed, it may not cut down appreciably on the number of yellow jackets seen near your home. Why not? Because yellow jackets forage far from their nests and many stings occur when a foraging yellow jacket is accidently (or intentionally) messed with. I have heard that dusting a foraging yellow jacket with a small handful of flour will cause it to head back to its nest to get cleaned up, thus revealing its nest site. You might want to try this.

■ **Nests in the ground.** To destroy a nest in the ground, spray a small amount of a 'Wasp-Freeze' type product (there are many)

Shellac is made out the excretion of female lac insects, a type of scale. ♦

after dark in and around the nest entry. Then, guessing how big the nest is, quickly spray 10 to 30 seconds of a pyrethrin-based product (such as **Tri-Die**) into the hole. Close the hole up with a stone. That's it. Be ready to run if something goes wrong.

'Wasp-Freeze' type products can be used alone, and are quite effective alone. But they do contain fluorinated hydrocarbons which destroy the ozone layer. Please use them sparingly.

If this is done on a cool, dark night the process goes very quickly with relatively little danger to the person treating the nest. Nevertheless, suitable protective gear should always be worn. Full protective gear would include coveralls, a beekeeper's hat and veil, and beekeeper's gloves or other sting-proof gloves. Improvise at your own risk.

By placing red cellophane over the light of your flashlight, you can work close to the nest without alarming the occupants.

■ **Aerial Nests.** Treating aerial nests is similar except the 'Wasp-Freeze' and then the pyrethrin product (such as **Tri-Die**) are sprayed in *and* onto the nest. Quickly, but calmly, cut or knock down the nest, and if possible place it in a plastic bag which can be disposed of in a safe place. Small umbrella wasp nests can be quickly crushed under foot.

■ **Traps.** Yellow jacket traps which contain food baits or pheromones as attractants are available which may, but usually do not, provide adequate control. These are more effective during the summer and should be used only around the perimeter of your property so as not to attract yellow jackets close to your home.

Oldtimers made a yellow jacket trap by hanging a fish or piece of liver from a tripod suspended over a bucket of water with a squirt of soap in it. The yellow jackets grab big chunks of flesh, fall in the water, and drown.

Another scale insect, found on cactus was the source of the famous red dye,

■ **Professional nest removers.** In a few communities, there are people who specialize in the non-chemical removal of yellow jacket nests at a nominal charge. The collected nests are then sold to pharmaceutical companies that manufacture anti-venom from the dead insects. These people use carbon dioxide to freeze and asphyxiate the yellow jackets along with a big dose of bravado. Occasional stings are an occupational hazard.

■ **Cold weather.** Remember that all yellow jacket nests will die by themselves after a hard frost or two. If the nest is not a direct threat, many times they can be left for nature to run its course. Also keep in mind that removing a single nest rarely cuts down on the number of foraging yellow jackets found around your home.

Even though the process is pretty simple, there is always the possibility of getting stung when working with yellow jackets. Even if things go wrong, you've still got to keep your wits about you to avoid serious harm. Be careful out there.

cochineal, considered the best "red" before synthetic dyes were invented. ♦ A

APPENDICES

OTHER PESTS

Of the almost one million species of insects that have been identified, only about 200 species in this country are what we usually consider household pests. In this book we have looked at the most common 20 or so species, but what about the other 180 species?

Most of these are different species from other parts of the country of the same common pests. Some are only occasionally indoor pests. A few, such as bedbugs, used to be more common, but are relatively rare today (due to better sanitation methods).

Below are some of the 'other pests' that you might encounter along with some brief suggestions for control measures. As always, your local cooperative extension service can help you identify pests whose name tags may have fallen off.

■ **Outside creatures (that occasionally wander in).** Sowbugs (potato bugs), centipedes, earwigs, springtails, slugs and snails, are all primarily outside pests that sometimes make pests of themselves indoors. These pests can often be controlled by keeping vegetation, debris, and bark mulch away from the house, using weather stripping around doors and windows, and lowering the moisture level around and in the house.

■ **Overwintering Pests.** Certain creatures look for warm, cozy places to spend the winter and make pests of themselves in the fall as they begin moving in and again in the spring on their way out. These include the harmless but annoying box elder bugs and elm leaf beetles, overwintering yellow jacket queens, and mice and possums and other furry creatures. The key to controlling these

harvester ant can lift 50 times its own weight. Some beetles can lift 850 times

seasonal pests is excluding them by tightening up the outside of your house with weather stripping and good screens on crawl space vents. Large numbers of pests can be vacuumed up daily until they subside.

■ **Pantry pests.** A whole slew of beetles and other insects occasionally attack stored goods including cadelles, cigarette beetles, drugstore beetles, grain weevils, and rice weevils. Most are relatively rare in homes, but in certain parts of the country may be more common. The key to controlling all pantry pests is to identify the infested food sources and get rid of them! Certain pests may require some spraying to help break up their life cycles.

■ **Wood Boring Insects.** Besides termites and carpenter ants discussed in this book, there are other wood destroying insects that show up occasionally. These include the wood boring beetles such as the powderpost beetles, the old house borer, and the beautiful metallic buprespid beetle which appears only after the damage has been done. Wood wasps or horntails and carpenter bees will also occasionally be found in or around the home.

Each of these pests requires its own control strategy. In general, wood destroying pests can be discouraged by keeping the area under and adjacent to your home dry, by storing firewood away from the house, and by removing wood debris from your property.

■ **Random Visitors.** Many insects, that are not truly household pests, show up in the home now and then, usually through open doors and windows. These include mosquitoes, night flying moths that are attracted to lights, honeybees, among the innumerable garden, field, and forest creatures that occasionally go astray. Random visiting is bound to happen in a universe governed by entropy, and can never be totally prevented. Again, the solution is to keep the critters out as best you can. Reducing the

their own weight. ♦ Some American Indians spoke of the honebee as "the white number of flowering plants near the house will cut down on unwanted bees. Try to enjoy these surprise visitors. Who knows when you will have a chance to see one again?!

CLOSE ENCOUNTERS OF THE FOURTH KIND — EXTERMINATORS

Despite my efforts to get you to 'do-it-yourself', it may be necessary at times to hire a professional exterminator. This can be an intimidating prospect as you may feel the need to protect yourself against both potentially harmful pesticides *and* potentially unscrupulous business people. Exterminators as a group are not really *that* bad, probably falling somewhere between dentists and the IRS. But, some are *much worse* than others, and these are the ones you have to look out for.

Here are a few tips to help you protect yourself from the bad apples. Even otherwise very competent people overlook the obvious when convinced that wood-destroying insects are rapidly destroying their homes.

- Be careful when you hear a high pressure sales pitch. There is never a need to sign a contract right there on the spot. There is no pest problem that cannot wait a few hours or days. The more feverish the pitch, the faster you should show the guy to the door.

- Think twice about working with companies offering 'free inspections'. Everyone in business is in business to make money. Odds are if they don't charge you up front, they'll make up for it somehow later, sometimes in a big way. Most companies use the offering of 'free inspections' as a way of generating sales leads. This is not inherently bad, but the temptation is great for salespeople to 'find' problems that aren't there, to avoid wasted trips.

man's fly" as honeybees were not native to North America. ♦ Honeybees can

● Always get several estimates, especially for bigger jobs. While the salesperson/inspector is visiting, make sure all your questions are answered to your satisfaction. This is not the time to be demur and retiring. Have him or her tell you exactly what chemicals and methods will be used. Make sure you are getting into something you can live with later.

As a rule, pest control people are not bad guys, but as an industry they have lagged behind in providing the type of service that you as a customer deserve. By being clear in expressing to them the kind of service you want, you help them know what kind of services they should be offering.

travel three miles to reach a nectar source. ◆ Monarch butterflies migrate many

LEAST TOXIC LAWN CARE

Many people feel that herbicides and insecticides are necessary evils when it comes to maintaining a beautiful and healthy lawn. Not so! Many lawn problems can be minimized or eliminated by following a well-conceived lawn maintenance program. The following are some tips to get you started. Try to find an enlightened garden supply store salesperson or Master Gardener to help you choose specific fertilizers and turf varieties for your area:

CULTURAL PRACTICES

- Avoid excess Nitrogen but maintain high Potassium levels. This helps avoid disease problems.

- Water correctly; not too little, not too much.

- Keep mowing blades sharp.

- Raise mowing height.

INSECTS

- Use resistant cultivars of grass seed or sod.

- Pheromone traps for Sod webworm, Armyworms and Cutworms.

- Predatory nematodes for soil inhabiting insects.

- Traps for moles and gophers.

WEEDS

- Pre-irrigate before seeding a new lawn. Then cultivate, use flamer, or hand pick weeds.

- Never allow weeds to go to seed.

- Use aggressive turf varieties to compete with weeds.

- Water through the summer.

hundreds of miles to their over-wintering places and back again. A monarch can

SOURCES OF PRODUCTS

The following list includes all products mentioned in this book and the addresses of their manufacturers or distributors. These products are ones that the author is familiar with and may not be any better than other similar products on the marketplace. Encourage your favorite stores to stock these or similar products.

PRODUCT NAME	DESCRIPTION	MANUFACTURER	DAN*
DEMIZE	flea spray with linalool	Pet Chemicals P.O. Box 17167 Memphis, TN 38187	+(+)
DIATOMACEOUS EARTH	desiccant for fleas	Pristine Products 2311 E. Indian Sch. Rd. Phoenix, AZ 85016	++
ELECTRO-GUN	non-chemical drywood termite control	Electro-gun 916 S. Casino Center Las Vegas, NV 89101	+++
FLEA COMB	specialized comb for pets	Save Our ecoSystem 541 Willamette St. Eugene, OR 97401	+++
FLEA STOP	citrus based flea dip	Pet Chemicals P.O. Box 17167 Memphis, TN 38187	++
HEAD LICE COMB	specialized comb for lice	JTLK Inc P.O. Box 427 Boonton, NJ 07005	+++
KETCHALL	multiple catch mouse trap	Kness Mfg. Co. P.O. Box 70 Albia, IA 52531	+++
MAGNETIC ROACH FOOD 2000	boric acid roach paste	Blue Diamond Mfg. P.O. Box 1322 Rogersville, TN 32857	++
PERMA-DUST	boric acid dust	Whitmire Labs 3568 Tree Ct. Indus. Blvd St. Louis, MO 63122	+
PRECOR	insect growth regulator	Zoecon 1220 Denton Dr. Dallas, TX 75234	+
REVENGE	pyrethrin/silica aerogel w/ injector	Roxide Intl. 3 Cottage Place New Rochelle, NY 10802	+

travel about 600 miles on one feeding of nectar. ♦ Harvester ants often have

ROACH-PRUFE	boric acid dust	Copper Brite Inc. P.O. Box 50610 Santa Barbara, CA 93150	++
SAFER'S INSECTICIDAL SOAP	soap-based insecticide	Safer Inc. 189 Well Avenue Wellesley, MA 02159	++
SNAP-E TRAP	'improved' snap-type mouse trap	Kness Mfg. P.O. Box 70 Albia, IA 52531	+++
SUREFIRE	pheromone traps	Concep Membranes P.O. Box 6059 Bend, OR 97708	+++
STIKEM SPECIAL	a 'glue' used for insect barriers	Seabright Enterprises 4026 Harlan St. Emeryville, CA 94608	++
TANGLEFOOT	a 'glue' used for insect barriers	The Tanglefoot Co. 314 Straight Ave. SW Grand Rapids, MI 49504	+++
TORUS 2E	fenoxycarb insect growth regulator	Maag Agrochemicals P.O. Box 6430 Vero Beach, FL 32961	+
TRI-DIE	pyrethrin/silica aerogel w/ injector	Whitmire Research Labs 3568 Tree Ct. Indus. Blvd St. Louis, MO 63122	+
VICTOR TIN CAT	multiple catch mouse trap	Woodstream Inc. P.O. Box 327 Lititz, PA 17543	+++
VICTOR HOLDFAST	roach sticky trap	Woodstream Inc. P.O. Box 327 Lititz, PA 17543	+++
XCLUDE	micro-encapsulated natural pyrethrum	Whitmire Research Labs 3568 Tree Ct. Indus. Blvd St. Louis, MO 63122	++

* The acronym DAN stands for Dan's (Arbitrary) Assessment of Negativity. The DAN rating is the author's attempt to give the reader a sense of the relative safety of the various products. Safety is a very slippery concept to measure. This scale is probably no less arbitrary or accurate than others from so-called experts. All products listed here are believed to be safe when used intelligently, but those with more 'pluses', are in Dan's opinion, probably 'more safe'.

small crickets in their nests which are tolerated like we tolerate pets in our own

FOR MORE INFORMATION...

A FEW GOOD BOOKS

Berry, Ralph E, *Insects and Mites of Economic Importance in the Northwest.* 1978. Corvallis, OR. OSU Book Stores.

Christensen, Chris, *Technician's Handbook to the Identification and Control of Insect Pests.* 1989. Cleveland, OH. Franzak and Foster Co.

Ebeling, Walter, *Urban Entomology.* 1975. Los Angeles, CA. Division of Agricultural Sciences, University of California.

Mallis, Arnold, *Handbook of Pest Control.* 1982. Cleveland, OH. Franzk and Foster Co.

ORGANIZATIONS

Without question, the single, most important source of 'least toxic' pest control information comes from the *Bio-Integral Resource Center.* Anyone seriously interested in this field should subscribe to one or both of their excellent journals: *The IPM Practitioner* (higher technical level for scientists); *Common Sense Pest Control Quarterly* (for the layperson). BIRC is also available for consulting on specific problems.

■ Bio-Integral Resource Center, P.O. Box 7414, Berkeley, CA 94707. (415) 524-2567

A good source for information on pesticides is the *Northwest Coalition for Alternatives to Pesticides (NCAP).* Their efforts concentrate on public education and on political issues associated with pesticides. NCAP is a membership organization and publishes the quarterly, *Journal of Pesticide Reform.*

■ Northwest Coalition for Alternatives to Pesticides, P.O. Box 1393, Eugene, OR 97440. (503) 344-5044

---

homes and Leafcutter ants sometimes share their nests with a tiny cockroach.

The author, Dan Stein is owner/operator of Northwest IPM, a full service pest control and consulting company in Eugene, Oregon, specializing in least toxic pest control.

Please write if you would like any of the following:

- A price list of supplies available from Northwest IPM.
- Brochure of services available from Northwest IPM.
- Information on Do-it-Yourself Carpenter Ant Kit.

Additional copies of this book ($10 postpaid) may be ordered from:

> Northwest IPM
> P.O. Box 11445,
> Eugene, OR 97440

Please direct library, bookstore and distributor inquiries to the publisher. Quantity discounts available.

> Hulogosi Communications, Inc.
> P.O. Box 1188
> Eugene, OR 97440

♦ "He prayeth best who loveth best all creatures great and small." ♦ THE END

TYPE SET BY HULOGOSI IN 11 PT. GAILLARD AND 9 PT. TEKTON.
PRINTED BY KOKE PRINTING COMPANY, EUGENE, OREGON.